45p

Periodicity and atomic structure

Longman ISBN 0 582 04351 4
Penguin ISBN 0 14 082.351 4
Nuffield Chemistry Special

Contents

Periodicity and atomic structure

N. N. Greenwood

Nuffield Chemistry Special
Published for the Nuffield Foundation by Longman/Penguin Books

The Longman Group, London and Harlow
Penguin Books Ltd, Harmondsworth, Middlesex

First published 1971

Filmset in 'Monophoto' Times New Roman by
Photoprint Plates Ltd, Rayleigh, Essex
and made and printed in Great Britain by
Compton Printing Ltd, London and Aylesbury

Designed by Ivan and Robin Dodd

Part One

Introduction

The sum total of chemical knowledge doubles every 10 years. If we consider everything that was discovered from the beginning of recorded history up to 1960, then this represents only half of what is known today. If we project forward from 1970 another 32 years to the year AD 2003 the amount known will have increased tenfold again (as a result of discoveries which many of you will have helped to make) and our present knowledge of chemistry will only be 10 per cent of what will then be known. This means that in the study of chemistry at all levels we must be concerned with concepts, relationships, and modes of thought rather than with lists of things to be learned. At the very centre of the interpretation of chemical phenomena stands the Periodic Table. If you have read the O-level Background Book *The Periodic Table* you will be aware of some of the experimental evidence for periodicity in physical properties and chemical behaviour. The word 'periodicity' means the recurrence among elements of certain features which share strong resemblances repeated in the group to which they belong. An example of this is the halogen group of elements. More evidence will be given at the end of Part Two, and in Parts Three and Four.

To understand why the chemical elements should show this periodicity we must examine the structure of the atom. We will then be able to appreciate why the Periodic Table has the form it does and how it can best be used for interpretations and prediction. Before considering the details of atomic structure however, it is worth pausing for a moment to reflect that every single object in the material universe is composed of atoms in various combinations but that there are only about 100 different types of atom. So this tremendous number of substances has grown from a few basic units. An understanding of this process has led to the recognition that the planets, the Sun, and the other stars contain precisely the same types of atom – one of the great generalizations of human knowledge and thought. It is possible to see how this diversity arises by considering atoms as analogous to the letters of the alphabet.

Language	Chemistry
alphabet	elements (atoms)
words	radicals, polyatomic ions
sentences	compounds (molecules, ionic crystals)
novels, poems	substances (mixtures)

Table 1
An analogy between language and chemistry

The infinite diversity of English literature can be expressed using 26 letters of the alphabet and a few punctuation marks. The 26 letters can be combined into some millions of different words and these, in turn, combined into an unlimited number of sentences. Finally, the sentences can be placed in sequence to make a novel, a poem, or a book like this one! So, in chemistry, the 104 types of atom can be combined into functional groups called radicals (or polyatomic ions if they are electrically charged) and these in turn can combine to give molecules which are the minute particles of pure compounds. Finally, these compounds can be mixed in a variety of ways to give the familiar substances of everyday life such as chalk, flesh, grass, plastic, and wood.

At present, over two million pure chemical compounds are known, and this number is increasing by some 100 000 compounds each year. Some method of organizing this vast array of factual material to make it coherent is obviously essential, and the Periodic Table helps us to do this.

We see that an understanding of chemistry depends ultimately on a knowledge of the structure of the atom and on how atoms combine with each other. But we do not need to appreciate the most modern theoretical refinements in order to understand the important features of atomic structure; once the broad outlines have been grasped, chemical periodicity and the Periodic Table become obvious.

Part Two

Atomic structure

Evidence leads us to believe that atoms have a complex internal structure, consisting of a positively charged nucleus surrounded by various sets of negatively charged electrons (see the Background Book *Inside the Atom*). The number of positive charges on the nucleus is called the atomic number, Z, and it determines what element is being considered and its place in the ordered sequence of elements. Thus, the element in which $Z = 1$ is called hydrogen, the element in which $Z = 2$ is called helium, and so on. The atomic number also tells us the number of electrons in the neutral atom; it is the number and arrangement of these electrons which give an atom its characteristic chemical properties. Let us look briefly at the properties of the nucleus and then go on to enquire about the detailed arrangement of the electrons around the nucleus.

The nucleus is composed of neutrons and protons bound very tightly together by the force field associated with still smaller particles called mesons. The neutron has unit mass and no charge. The proton has unit mass and unit positive charge. It is, in fact, the positive hydrogen ion, H^+. At a first approximation therefore, the mass of a nucleus is given by the total number of neutrons and protons it contains, whereas the atomic number is determined only by the number of protons. For many of the lighter elements in the Periodic Table, the atoms contain approximately equal numbers of neutrons and protons but it is often possible for an element to have atoms in which the number of neutrons varies. Such atoms are called isotopes. Some examples of isotopes are given in the table.

Element	Protons	Neutrons	Nuclear mass	Symbol
Hydrogen	1	0	1	$^{1}_{1}H$
	1	1	2	$^{2}_{1}H$ (deuterium)
	1	2	3*	$^{3}_{1}H$* (tritium)
Helium	2	1	3	$^{3}_{2}He$
	2	2	4	$^{4}_{2}He$ (α particle)
Carbon	6	6	12	$^{12}_{6}C$
	6	7	13	$^{13}_{6}C$
	6	8	14*	$^{14}_{6}C$* (radio-carbon)
Potassium	19	20	39	$^{39}_{20}K$
	19	21	40*	$^{40}_{20}K$*
	19	22	41	$^{41}_{20}K$

Table 2

The constitution of some isotopes—*indicates that the isotope is radioactive

The existence of isotopes explains why many elements have atomic weights which vary considerably from whole numbers; this is because the atomic weight is the average weight of all the isotopes after the abundance of each has been taken into account. For example, boron has two stable isotopes, of mass 10 and 11; the relative abundance of these isotopes is approximately of the ratio 1:4, hence the atomic weight of boron is:

$$\frac{(1 \times 10)+(4 \times 11)}{(1+4)} = 10.8.$$

It is not surprising that the relative abundance of the isotopes of several elements varies from place to place and that this results in a slight variation in the atomic weights of these elements. Boron, for example, varies between 10.808 and 10.814.

If the nucleus contains too many or too few neutrons for stability, then the atom is radioactive and the nucleus rearranges itself into a more stable configuration, frequently having a different number of protons. Thus, the nucleus of potassium-40 can rearrange itself by capturing one of the extra-nuclear electrons and in this way transforming itself into an ion of argon which rapidly picks up an electron from the surrounding material to become a neutral atom. The nucleus of potassium-40 can also achieve stability by ejecting a high speed electron called a beta (β) particle and transforming itself into the nucleus of the more stable calcium-40. Half the potassium-40 in the world is converted into argon-40 or calcium-40 every 1270 million years.

$$^{40}_{19}\mathrm{K} \xrightarrow{\text{electron capture}} {}^{40}_{18}\mathrm{Ar}$$

$$^{40}_{19}\mathrm{K} \xrightarrow{\beta \text{ particle}} {}^{40}_{20}\mathrm{Ca}$$

This process immediately explains why argon is very much more abundant in the atmosphere than any of the other inert gases, because it is being continuously generated by the much more abundant element potassium. It also explains why the atomic weight of argon is greater than that of potassium although its atomic number is less. (See discussion on page 22.) One more thing should be mentioned about the nucleus – its small size. Its diameter is only about $\frac{1}{100\,000}$ the diameter of the whole atom and it is worth while to reflect on the tremendous binding power of nuclear forces which enable them to keep so many positively charged protons so close together. Nuclear forces are about 100 times stronger than electrostatic or chemical bonding forces but, because they operate over distances of only about 10^{-13} centimetre, the nuclear energies associated with them are some five million times greater than chemical binding energies which operate over distances of about 10^{-8} centimetre.

Some of this tremendous energy can be evolved during nuclear fission and fusion reactions.

Extra-nuclear electrons – The charge on an electron is equal in size, and opposite in sign, to that on a proton but measurements of the ratio of charge to mass for the two particles show that the electron has only $\frac{1}{1836}$ the mass of the proton. It follows that the mass of the atom is concentrated almost entirely within the nucleus and that for electrical neutrality the number of electrons equals the atomic number.

The rules describing the behaviour of electrons appear rather strange at first because it is difficult to find an analogy for them in the large-scale world of everyday observation. Because electrons are so small and weigh so little it is impossible to observe them in any experiment without disturbing them. If an experiment is devised to give precise information about the position of an electron, it is found that very little can be determined simultaneously about its velocity. This also applies the other way round: you can find out precisely about the electron's velocity but not at the same time about its position. These ideas were developed by the German physicist, Werner Heisenberg, in his Uncertainty Principle (1927) and they force us to adopt a statistical interpretation of atomic phenomena. We can only talk about a *probability* that an

Figure 1
Professor W. Heisenberg (b. 1901) who developed the Uncertainty Principle.

electron will be in a certain place – it is as though the electron, instead of being a point charge, were smeared out into a cloud or distribution of charge. The appropriate mathematical description is the same as that used to describe waves and is sometimes called wave mechanics. The full theory cannot be developed here but the results turn out to be quite simple to state and quite easy to understand. They are set out in the following paragraphs.

The four quantum numbers – The arrangement of electrons in an atom can be described by means of four variables which are called quantum numbers because they can only take on certain definite values. It is this discontinuous or quantum nature of the variables which distinguishes the description of atomic phenomena from the more familiar ways in which we describe macroscopic phenomena. For example, the state of a gas can be described by three variables – pressure, temperature, and volume – and these can have continuously changing values. By contrast, the quantum numbers describing electrons in atoms can only take on certain definite (usually whole number) values. Let us look more closely at these quantum numbers.

The first, n, is called the *principal quantum number*, and it defines the general energy level or 'shell' to which the electron belongs. It is important to realize, however, that these are not shells in space, like the layers of an onion, but shells in energy. Electrons with $n = 1$ are the most tightly bound. Then come another group of electrons, much less firmly bound, for which $n = 2$, and so on. The experimental evidence for these discontinuous energy levels comes from ionization energies (figure 7). We shall also find that the outermost electrons for a given atom have a principal quantum number equal to the number of the period (horizontal row) in which the atom is to be found in the Periodic Table. (Electrons with n equal to 1, 2, 3, 4, and so on, are sometimes said to belong to the K, L, M, N ... shells of the atom.)

The second quantum number, l, is called the *orbital quantum number* and defines both the shape of the electron charge distribution and its orbital angular momentum. The number of different shapes, however, is limited and is given by the principal quantum number, n. An electron with $n = 1$ can only be spherically symmetrical that is, there is an equal probability of finding the electron in any direction from the nucleus. This spherical distribution of its charge has zero orbital angular momentum and l is given the value 0.

When $n = 2$, the orbital quantum number, l, can have one of two values, 0 or 1. As before, $l = 0$ refers to a spherical charge distribution with zero orbital angular momentum, but $l = 1$ gives a dumb-bell-like shape which is associated with a definite angular momentum.

Similarly, when $n = 3$, l can have the values 0 and 1 as before and also a third value $l = 2$ for which the shape of the charge distribution is even more complex.

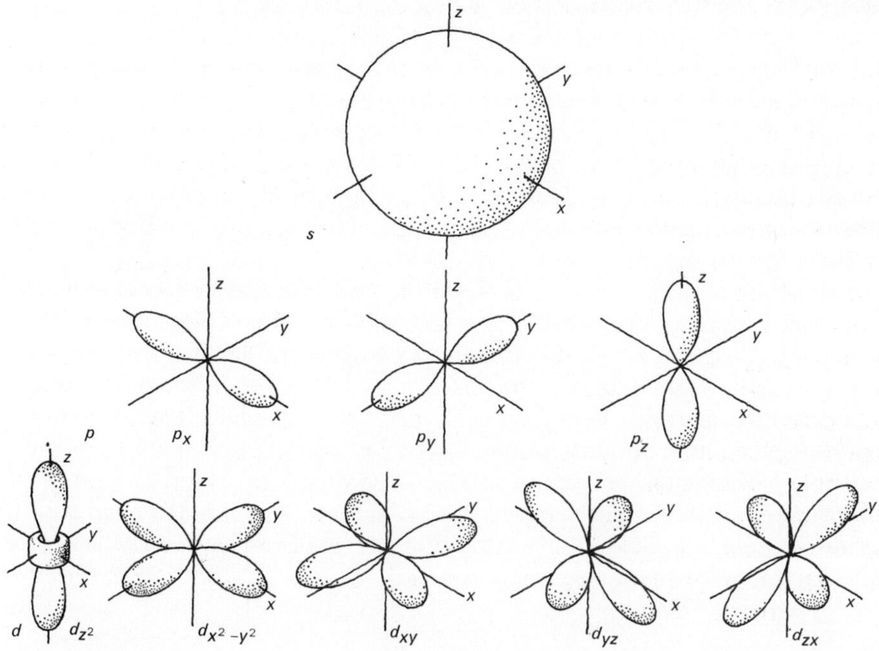

Figure 2
Approximate indication of the shape of the charge distribution for *s*, *p*, and *d* electrons.

In discussing electron distributions it is convenient to use a set of symbols which indicates the principal quantum number and orbital quantum number of each electron. This is done by using the actual numbers, 1, 2, 3, and so on, for the principal quantum number and the abbreviations, *s*, *p*, *d*, and *f* for electrons having $l = 0$, 1, 2, and 3 respectively. At first sight, these symbols may seem to be chosen at random but they arose historically from the nomenclature of atomic spectroscopy from which the spacings of the various atomic energy levels are derived. In these spectra it is possible to identify various series of lines called sharp, principal, diffuse, and fundamental, and the initial letters of these series have been used to label the electrons involved. Here are a few examples.

a 4*s* electron has $n = 4$, $l = 0$
a 2*p* electron has $n = 2$, $l = 1$
a 3*d* electron has $n = 3$, $l = 2$
a 5*f* electron has $n = 5$, $l = 3$

It is most important to get some practice in using this nomenclature because it is employed throughout chemistry. We shall also find that each type of electron, *s*, *p*, *d*, or *f*, is associated with a definite area of the Periodic Table.

So far in our treatment we have mentioned two quantum numbers, the principal quantum number, *n*, which defines the general energy level to which the electron belongs, and the orbital quantum number, *l*, which specifies the shape and orbital angular momentum of the electron. Now a further question arises: if the charge distribution of the electron is not spherically symmetrical, in what direction is it oriented in space? This brings us to the third quantum number.

The third quantum number, *m*, is called the *magnetic quantum number* for it is only in an applied magnetic field that it is possible to define a direction within the atom with respect to which the orbital can be oriented. The magnetic quantum number can have any integral value from $-l$ to $+l$ and each value corresponds to a definite orientation with respect to the applied magnetic field. We can see again that quantum conditions apply strongly: not all alignments are possible but only certain well defined ones. This is because the orbital motion of the electron generates a small magnetic field of its own and this can align with or against the external field, or at various definite angles in between which correspond to separate quanta of energy. A typical set of orientations for a *d* electron ($l = 2$) is shown in figure 3.

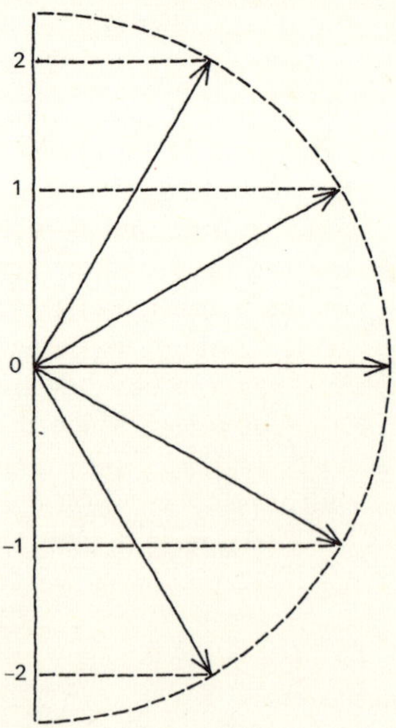

Figure 3
Magnetic quantum numbers (m) for d orbitals ($l = 2$).

In general the magnetic quantum number can take up $2l+1$ values. From this we see that there can be only one orientation for an *s* orbital (because it is spherical and has zero orbital angular momentum) but that a *p* orbital can have three orientations; these orientations can be conveniently chosen as the *x*, *y*, and *z* directions in Cartesian co-ordinates. Likewise, there are five possibilities for *d* orbitals and seven for *f* orbitals.

The fourth and final quantum number, m_s, is called the *spin angular momentum quantum number*. To discuss this concept properly in the language of wave mechanics requires the introduction of relativity (four-dimensional space–time) but fortunately for chemists, this is not necessary and we can revert to the language of particles to describe what is going on. As the name 'spin quantum number' implies, we imagine that the electron is spinning about its own axis and that this motion is associated with a magnetic field. The spin angular momentum can be coupled either with or against the orbital angular momentum so that each configuration is split into two sub-states. The two directions of spin are often designated $+\frac{1}{2}$ and $-\frac{1}{2}$ so that in changing from one direction to the other the spin quantum number changes by one unit as required by the quantum theory.

By means of the four quantum numbers we are now able to specify the configuration of each extra-nuclear electron precisely, by indicating its energy level (n), its shape and orbital angular momentum (l), its orientation (m), and its spin angular momentum (m_s). In short, we have described the ingredients of the extra-nuclear part of the atom but to put these ingredients together correctly we need a recipe. This is supplied by the three rules or principles which are set out in the next section.

Building up the atom – Imagine that we have a bare nucleus of positive charge Z; to make a neutral atom, Z electrons must be added. The overall electron distribution obtained must be such that the atom is in the state of lowest possible potential energy. This is the first rule and it merely states that each added electron should be placed in the most strongly bound position. (An atom in this state is said to be in its ground state.)

The second rule gives the sequence of the various electron shells in order of increasing energy (decreasing tightness of binding). For elements of low atomic number the rule states that: the energy sequence follows the sequence of values of the sum $(n+l)$; if two orbitals have the same value of $(n+l)$ then the one with lower n has the lower energy and fills first. This means that, as we have already seen, the energy is primarily determined by the value of n but that it is modified to some extent by the value of l. The sequence of levels which this rule generates is shown in figure 4 and will be discussed more fully in Part Three.

Orbital	*n*	*l*	*n*+*l*
1*s*	1	0	1
2*s*	2	0	2
2*p*	2	1	3
3*s*	3	0	3
3*p*	3	1	4
4*s*	4	0	4
3*d*	3	2	5
4*p*	4	1	5
5*s*	5	0	5
4*d*	4	2	6
5*p*	5	1	6
6*s*	6	0	6
4*f*	4	3	7
5*d*	5	2	7
6*p*	6	1	7
7*s*	7	0	7
5*f*	5	3	8
6*d*	6	2	8

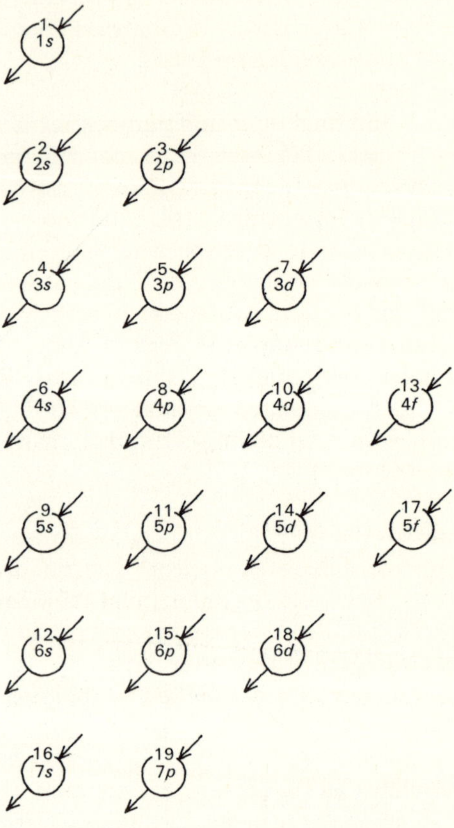

Figure 4
The sequence of energy levels.

For elements of high atomic number the principal quantum number becomes even more predominant and the rule alters slightly. For these elements the energy sequence tends to follow the values of *n*, and for each *n* value the sequence follows the increasing values of *l*. The sequence therefore becomes 1*s*; 2*s*, 2*p*; 3*s*, 3*p*, 3*d*; 4*s*, 4*p*, 4*d*, 4*f*; and so on.

The third rule tells us how many electrons can be put in each energy level. It is called the Pauli Exclusion Principle (after the Austrian scientist Wolfgang Pauli) and for our purpose it may be stated as follows: each electron in an atom

Figure 5
Niels Bohr (1885–1968).
Photo, Lotte Meitner-Graf

must have a unique set of the four quantum numbers – no two electrons can have the same values for all four quantum numbers. For example, if a 1*s* electron has $m_s = +\frac{1}{2}$ then no other electron in that atom can have $n = 1, l = 0, m = 0$, and $m_s = +\frac{1}{2}$, since these four values for the quantum number have already been pre-empted by the first electron.

Experimental evidence – The existence of these various electronic energy levels within an atom is strikingly demonstrated by the spectra of light emitted from excited atoms when they return to states of lower energy and finally to the ground state. This light does not cover a continuously varying range of frequencies like the spectrum of a rainbow; it occurs only at certain well-defined frequencies which, as the Danish physicist Niels Bohr first showed, are related to the difference between the various energy levels in the atom. The spectrum for atomic hydrogen is particularly simple and part of it is shown in figure 6.

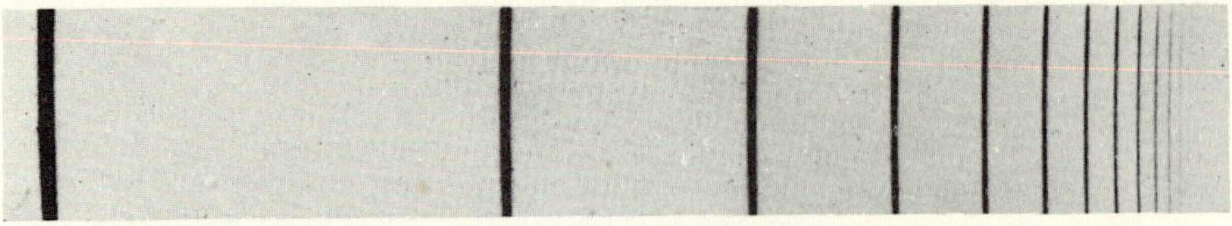

Figure 6
Spectrum of hydrogen.
From Herzberg, G. (1927) Annalen der Physik **84**

In this region of the spectrum each line represents a transition from some higher energy level to the level with $n = 2$. Other sections of the hydrogen spectrum occur in the ultraviolet region (transitions to $n = 1$) and in the infra-red (transitions to $n = 3$, $n = 4$, etc.). For heavier atoms the spectra are more complex but give us similar information. The detailed interpretation of the atomic spectra of hydrogen and other elements need not detain us; we merely note the clear evidence they provide for discrete electronic energy levels in atoms.

Other experimental evidence for the existence of electronic energy levels and shells in atoms comes from the measurement of ionization energies. The ionization energy of an electron in an atom or ion is the energy required to remove the electron from the free atom or ion to infinity. The experimental data for the twelve electrons in magnesium ($Z = 12$) are given in figure 7

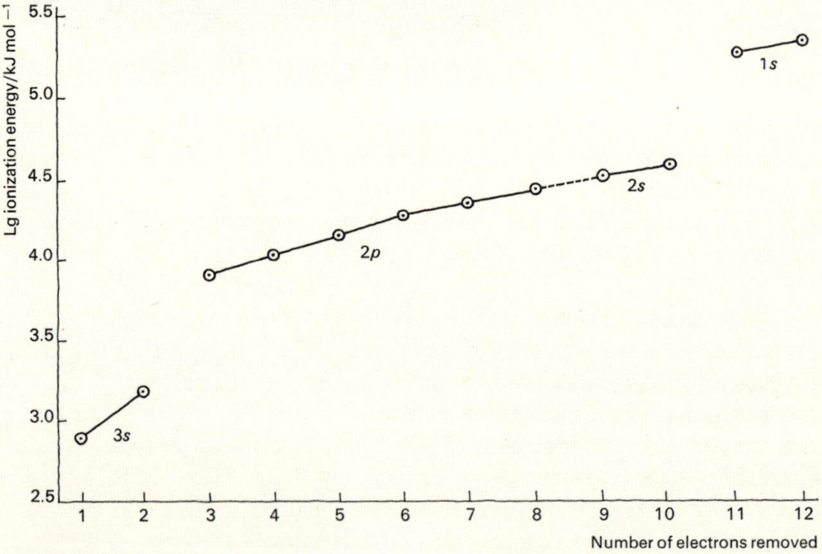

Figure 7
Successive ionization energies of magnesium.

which plots the logarithm of the ionization energy in kilojoules per mole against the number of the electron removed.

A logarithmic plot is used because it requires several hundred times more energy to remove the twelfth electron ($Mg^{11+} \longrightarrow Mg^{12+} + e^-$) than it does to remove the first ($Mg \longrightarrow Mg^+ + e^-$). The diagram shows that the first two electrons are much more loosely bound than the next eight and these in turn are more readily removed than the innermost shell of two electrons. Within every shell each electron has a well-defined binding energy as required by the quantum theory.

On the basis of this experimental evidence and the theoretical framework outlined earlier in this Part we are now able to relate atomic structure to the Periodic Table. The next Part shows just how close this relation is.

Part Three

The Periodic Table

Let us apply the quantum theory of atomic structure and the Pauli exclusion principle to each element in turn beginning with hydrogen, the lightest. This Part will get increasingly difficult but go on reading carefully as far as you can. It does not greatly matter from the point of view of understanding the rest of this book if you do not finish this Part on the first reading.

First row – The first element, hydrogen, has atomic number $Z = 1$ and therefore it has one electron surrounding the nucleus. In its ground state this electron will have $n = 1$, $l = 0$, and must therefore also have $m = 0$. It is thus a 1*s* electron and has spin $+\frac{1}{2}$. The second element, helium, has two electrons; both can go in the 1*s* orbital providing their spins are opposed, for in this way they can avoid having all four quantum numbers identical with each other. The electron configuration of helium is thus $1s^2$. There is no other possible combination for $n = 1$ because l can only have one value in this case. Hence the first energy level is full.

The ionization energy of atomic hydrogen (that is, the energy required to remove the electron from the atom leaving a bare proton) is 1310 kJ mol^{-1}. The energy required to remove one electron from helium is 2370 kJ mol^{-1}; this is much larger because the charge on the helium nucleus is twice that on the hydrogen nucleus and is not completely shielded by the second electron. This very high ionization energy from helium means that helium does not form compounds in which it exists as a positive ion and also makes it extremely unlikely that covalent chemical compounds will be formed. Helium is therefore expected to be an inert, monatomic gas, which is what it turns out to be.

Second row – For $n = 2$, l can have two values 0 and 1, and for $l = 1$, m can be $+1$, 0, or -1. Each combination of n, l, and m, can be associated with two electrons with opposed spins as set out below.

n	l	m	m_s	
2	0	0	$+\frac{1}{2}$	2 electrons
2	0	0	$-\frac{1}{2}$	

n	l	m	m_s	
2	1	+1	$+\frac{1}{2}$	6 electrons
2	1	+1	$-\frac{1}{2}$	
2	1	0	$+\frac{1}{2}$	
2	1	0	$-\frac{1}{2}$	
2	1	−1	$+\frac{1}{2}$	
2	1	−1	$-\frac{1}{2}$	

No other combinations based on $n = 2$ are possible. We know from figure 4 that the 2*s* orbital has lower energy than the 2*p*, and so will fill first. The electron configurations of the next two elements, lithium and beryllium, are thus Li $1s^2 2s$ and Be $1s^2 2s^2$. The next six elements gradually fill the 2*p* levels, at first singly so that the electrons can keep as far apart from each other as possible to reduce electrostatic repulsion, and then doubling up with opposed spins. The process is pictorially represented in the diagram, where \ represents $m_s = -\frac{1}{2}$ and / represents $m_s = +\frac{1}{2}$. The second row is now complete and this coincides with the completion of the second energy level ($n = 2$). This second row of elements in the Periodic Table is often called the first short period.

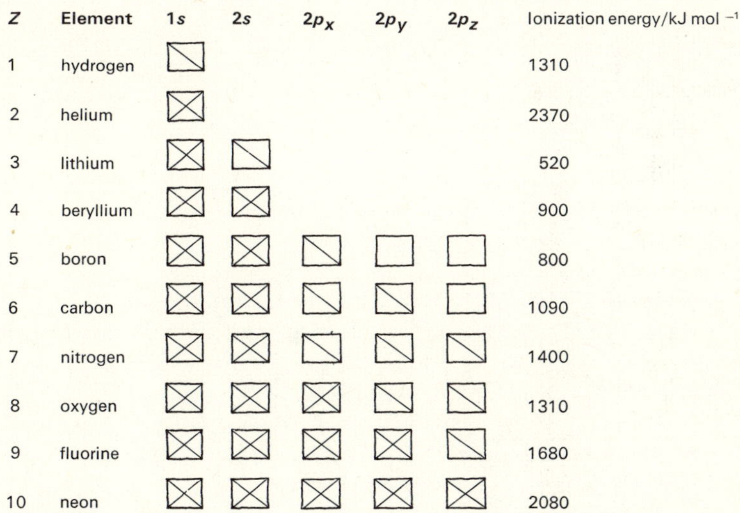

Figure 8
Electronic structure and ionization energies for the first ten elements in the Periodic Table.

The experimental values for the first ionization energies of these elements repay careful study because they parallel most convincingly the theoretical expectations. Thus, there is a large drop between helium and lithium as the principal quantum number n increases from 1 to 2, and then the ionization energy value rises somewhat for beryllium, though not to a value which is so high that beryllium would be expected to be an inert gas. Another decrease in ionization energy occurs at boron as the orbital quantum number, l, increases from 0 to 1, but this decrease is smaller than that between helium and lithium since an increase in l affects the binding energy less than does an increase in n. Notice also the very slight decrease between nitrogen and oxygen as the last three p electrons are added against the coulombic repulsion of the three p electrons already present, thus reducing the binding energy.

It can be seen that the ionization energies of the first ten elements are entirely consistent with our ideas of atomic structure, based on the quantum theory and the Pauli exclusion principle. Figure 9 illustrates well the successive increases and decreases of ionization energy discussed in the preceding paragraph.

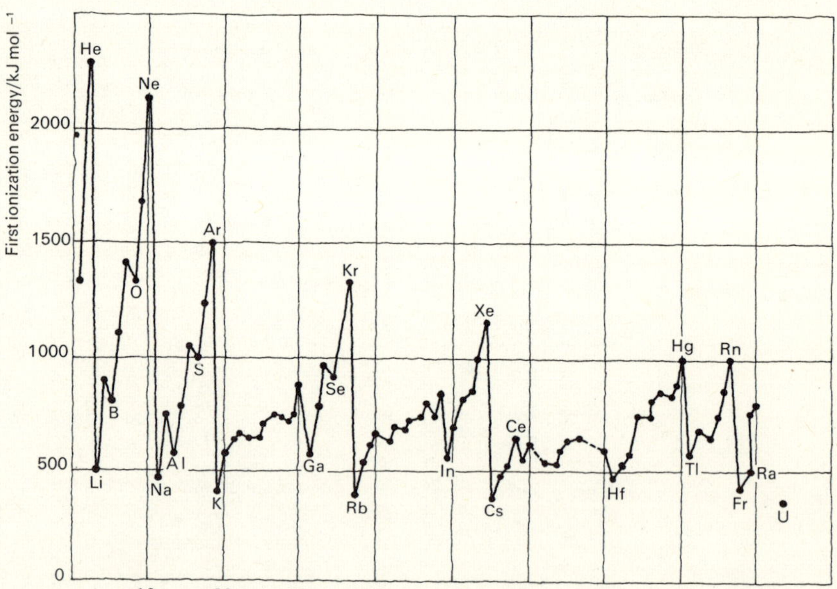

Figure 9
First ionization energies of the elements.

Third row – The process of filling up the energy levels is repeated for the second short period of elements from $Z = 11$ (sodium) to $Z = 18$ (argon). The 3*s* level takes two electrons and the 3*p* level six electrons as required by the theory and observed experimentally in the figure. At argon, the electron configuration is $1s^2$; $2s^22p^6$; $3s^23p^6$, and reference to figure 4 shows that the next available level is 4*s* ($n+l = 4$) rather than 3*d* ($n+l = 5$). We see that the completion of the second short period does not coincide with the complete filling of the third energy level ($n = 3$). There is, nevertheless, a distinct discontinuity in energy levels at this point as seen from the plot of ionization energies in figure 9; this is because, for most of the time, the 4*s* electron is further away from the positive nucleus than are the 3*s* and 3*p* electrons.

Already the patterns of the Periodic Table are beginning to emerge. The alkali metals have an outermost electron configuration ns^1 where *n*, the principal quantum number, is also the number of the row in the Periodic Table. These metals fall in the troughs of the ionization energy curve in figure 9 as their electrons are most easily removed. They are therefore highly reactive and give characteristic flame colorations when the loosely bound electrons are excited thermally. Likewise, the alkaline earth metals have the configuration ns^2 and are somewhat less reactive. Then there is a block six elements wide in which the *np* orbitals are filling, ending in each case with an inert gas having the configuration ns^2np^6. The inert gases occur at the peaks of the ionization energy curve and are the least reactive of the chemical elements. Note, however, that the values for xenon and radon are similar to that for mercury so that compounds of these elements might be expected; in fact many stable compounds of xenon have been made during the last few years and stable compounds of radon and even krypton have been prepared.

Fourth row: first long period – The nineteenth and twentieth electrons in element 19 (potassium) and element 20 (calcium) enter the 4*s* level. The next available group of energy levels is the 3*d* and, as $l = 2$ for these electrons, it follows that there are five *d* sub-levels corresponding to $m = +2, +1, 0, -1$, and -2 respectively. For each of these, m_s can be $+\frac{1}{2}$ or $-\frac{1}{2}$ so there is room for 10 electrons. These levels are filled in a fairly regular manner in the next ten elements and we note from the Periodic Table that it is just at this point that the first set of transition elements occurs from scandium $3d^14s^2$ to zinc $3d^{10}4s^2$. The fact that an inner level is filling in this region is reflected in the ionization energy curve (figure 9) which shows less variation than when *s* or *p* levels are being filled. The chemistry of these elements is also characteristically different from those of the alkali metals and alkaline earths (which involve *s* electrons) and the predominantly non-metallic group of elements in which the *p* electron levels are being completed.

After zinc there follow a further six elements in which the $4p$ levels are filled and this sequence ends with the inert gas krypton, $3d^{10}4s^2 4p^6$. We see that the expansion of the length of the period from eight elements to eighteen is due to the filling of the $3d$ levels in this region of the table.

Fifth row: second long period – After krypton ($Z = 36$) our rules would lead us to expect the successive filling of the $5s$, $4d$, and $5p$ levels with two, ten, and six electrons respectively. This is just what happens in rubidium and strontium, the second transition series from yttrium to cadmium, and finally the six elements ending with tellurium, iodine, and xenon. The physical properties and chemical behaviour of each of these eighteen elements resemble those of the corresponding member in the preceding long period and we conclude that periodicity in properties is indeed related to periodicity in atomic structure.

At this point it is salutary to realize that tellurium, iodine, and xenon ($Z = 52$, 53, 54 respectively) are only halfway down the list of 104 elements. It is customary to regard these three elements as the heaviest 'common' members of Groups VI, VII, and VIII respectively. But fifty other elements follow them in the Periodic Table including some of considerable industrial importance such as tungsten, platinum, gold, mercury, lead, and uranium. The patterns of electron energy level filling for the earlier elements still apply to these heavier elements and the only new phenomenon is the advent of f electrons and the inner transition series.

Heavy elements – After xenon ($5p^6$) the next most stable level is $6s$. An alkali metal (caesium) and an alkaline earth metal (barium) are therefore expected. These are followed by one transition element, lanthanum ($6s^2 5d^1$) after which the $4f$ levels become available.

We saw earlier that f electrons are ones for which the orbital quantum number, l, is 3. Seven values of the magnetic quantum number, m, are therefore possible: $+3$, $+2$, $+1$, 0, -1, -2, and -3. For each of these the spin quantum number, m_s, can be $+\frac{1}{2}$, or $-\frac{1}{2}$ so that fourteen places should be available. These places are occupied by the rare earth or lanthanide elements. This prediction, made by Bohr in 1913, was particularly valuable for two reasons. First, the lanthanides resemble each other very closely because a well-shielded inner level is being filled, and this similarity makes separation, purification, and identification difficult. During the nineteenth century, more than seventy unsubstantiated claims for new elements were made in this region of the Periodic Table. The certainty that no more than fourteen elements could exist between lanthanum and hafnium (an element also predicted by Bohr, and discovered in 1923) did much to clear the confusion in this field. Second, it became clear that there was an element missing between neodymium ($Z = 60$) and samarium ($Z = 62$). This turned out to be one of the lanthanides, promethium, which does not exist at all in nature since all its isotopes are radioactive with short half lives. Promethium was first isolated in 1945 from the fission products of uranium and since then, eleven isotopes have been synthesized.

The third transition series (5*d*) continues after this interruption and is complete at mercury, $4f^{14}$; $5s^2 5p^6 5d^{10}$; $6s^2$. The expected six elements follow and the 6*p* levels are completed by the inert gas radon. The 7*s* orbital is occupied by francium ($Z = 87$) and radium ($Z = 88$) and the fourth transition series (6*d*) begins with actinium ($Z = 89$). Unlike the 5*d* series, however, which was immediately superseded after lanthanum by the fourteen 4*f* lanthanide elements, actinium is followed by thorium $6d^2 7s^2$, and it is only in subsequent elements that the 5*f* levels become occupied: for example, protactinium, $5f^2 6d^1 7s^2$ and uranium, $5f^3 6d^1 7s^2$.

The elements after uranium are man-made. They do not occur in nature and become increasingly unstable towards α particle emission, electron capture, and spontaneous fission. For example, the longest lived isotope of uranium has a half life of 4.5×10^9 years whereas for neptunium it is 2.2×10^6 years and for americium 8.0×10^3 years. By the time californium, einsteinium, and fermium are reached, the longest lived isotopes have half lives of only 800 years, 250 days, and 11 days, and the most stable isotopes of mendelevium and lawrencium a mere 1.5 hours and 8 seconds. It is unlikely that an extensive chemistry of these elements will develop: nevertheless, what *is* known about their properties is consistent with their formulation as a second inner transition series of fourteen elements in which the 5*f* levels are gradually filled. There is some indication that the nuclei of elements beyond about atomic number 104 or 105 increase stability again and these elements are expected to continue the interrupted filling of the 6*d* levels.

Part Four

Periodicity in properties

In Part Three we saw the intimate relation between atomic structure and the Periodic Table. Let us now look back briefly to see how, for a hundred years between 1815 and 1913, chemists were moving towards this great generalization. The important point is that, as electronic structure is periodic (in the sense that certain configurations of *s*, *p*, *d*, and *f* electrons repeat themselves with increasing atomic number) so anything that depends on electronic structure will reflect this same periodicity to a greater or lesser extent. With the wisdom of hindsight we are therefore not surprised that many chemists noted regularities and gradations in properties of the elements and were slowly able to group similar elements together.

Early classifications of the elements – Prout in 1815 noticed that many atomic weights were direct multiples of the atomic weight of hydrogen; we see this as a consequence of the similarity in mass of the proton and neutron and the comparatively negligible mass of the electron. Exceptions to Prout's hypothesis point to the existence of isotopic mixtures.

Döbereiner in 1817 noticed that the properties of calcium, strontium, and barium were similar, with strontium intermediate between the other two. He later noticed other triads – chlorine, bromine, and iodine; lithium, sodium, and potassium; and so on. The modern interpretation is that the members of each triad occur after regular increments of 8 or 18 in the atomic number so that each triad of elements has a similar outer electronic configuration:

$$s^1 \text{ (Li, Na, K)}; \quad s^2 \text{ (Ca, Sr, Ba)}; \quad s^2p^4 \text{ (S, Se, Te)}; \quad s^2p^5 \text{ (Cl, Br, I)}.$$

Further progress depended on the provision of a reliable basis for the determination of atomic weights. Cannizzaro pointed out in 1860 that Avogadro's hypothesis provided such a basis. A more extensive classification of the elements, the 'telluric' screw of de Chancourtois soon appeared (in 1862) and this was followed (in 1864) by Newlands's 'law of octaves'. (See the Background Book *The Periodic Table.*) It was not only the musical analogy which told against the immediate acceptance of Newlands's periodic classification but also the fact that it broke down after two octaves and required the positions of iodine and tellurium to be inverted. We now know that after the second short period there is an expansion due to the ten *d* electrons in the transition elements and we also realize that, had the inert gases been discovered thirty-five years earlier, it would have been a law of nonaves instead of octaves.

Simultaneously with Newlands's work, rapid advances were being made independently by two other chemists. Meyer plotted atomic volumes against atomic weights and noticed their periodicity; the light alkali metals occur at the peaks of this curve and the transition elements, which are much denser, are found in the troughs. Mendeleev tabulated the elements according to their atomic weights (figure 10) but left gaps for elements not yet discovered (for example Sc, Ga, Ge) in places where the chemical properties of known elements enabled him to align similar elements in the same group. He also listed certain elements as a separate group outside the main table: Fe, Co, Ni; Ru, Rh, Pd; Os, Ir, Pt. A glance at the modern Periodic Table shows the reason for this – Mendeleev was trying to fit ten elements into the space of seven, which left the groups of three outside the table. We now recognize the ten elements in each period as the transition metals and the seven elements as the main groups but without the inert gases.

Series	Group I — R_2O	Group II — RO	Group III — R_2O_3	Group IV RH_4 RO_2	Group V RH_3 R_2O_5	Group VI RH_2 RO_3	Group VII RH R_2O_7	Group VIII — RO_4
1	H=1							
2	Li=7	Be=9.4	B=11	C=12	N=14	O=16	F=19	
3	Na=23	Mg=24	Al=27.3	Si=28	P=31	S=32	Cl=35.5	
4	K=39	Ca=40	– =44	Ti=48	V=51	Cr=52	Mn=55	Fe=56, Co=59 Ni=59, Cu=63
5	(Cu=63)	Zn=65	– =68	– =72	As=75	Se=78	Br=80	
6	Rb=85	Sr=87	?Yt=88	Zr=90	Nb=94	Mo=96	– =100	Ru=104, Rh=104 Pd=106, Ag=108
7	(Ag=108)	Cd=112	In=113	Sn=118	Sb=122	Te=125?	I=127	– – – –
8	Cs=133	Ba=137	?Di=138	?Ce=140	—	—	—	
9	(–)	–	–	–	–	–	–	
10	–	–	?Er=178	?La=180	Ta=182	W=184	–	Os=195, Ir=197 Pt=198, Au=199
11	(Au=199)	Hg=200	Tl=204	Pb=207	Bi=208	–	–	
12	–	–	–	Th=231	–	U=240	–	– – –

Figure 10
This Periodic Table, produced by Mendeleev in 1872, is in principle the same as our modern one. It contains many gaps for elements which Mendeleev predicted but which had not then been discovered.

The actual geometric form of the Periodic Table used by Mendeleev is not used today because it tends to overemphasize vertical similarities and to obscure the continuous gradation of properties across each period. It is also unable to handle the lanthanide (4*f*) elements convincingly. The great advance made by Bohr in 1913 was to indicate the underlying reason for periodicity in terms of atomic structure and to specify how many elements were to be expected in each period. Subsequently, wave mechanics gave an even deeper insight into this correlation.

The modern form of the Periodic Table is essentially a rearrangement of Mendeleev's table and emphasizes the relation of elements by atomic structure by placing them together, according to the *s*, *p*, *d*, and *f* groups of electron levels. The sequence of elements also follows the atomic number rather than the atomic weight, and this rationalizes the inversions of Ar and K, Co and Ni, and Te and I, which were adopted initially for purely chemical reasons.

In the two decades after its publication by Mendeleev the Periodic Table was extensively used to correct atomic weights, to predict new elements, and to suggest the most likely ways of discovering and isolating these new elements. It also found increasing use as a way of tidying up and making systematic a vast array of factual chemical knowledge, and some indication of this last use will now be given. Physical properties will be considered as well as chemical properties because there is a gradation between these two and the distinction is often quite arbitrary.

Periodic variation in properties

The ionization energy of an element has already been shown to reflect atomic structure particularly closely. It is instructive to colour in a blank Periodic Table like a contour map in physical geography. Use different colours for the following ranges of ionization energy: >1500, 1000–1500, 800–1000, 650–800, 500–650, <500 kJ mol^{-1}. What conclusions can you draw? More subtle variations also exist. Elements immediately following a closed shell or half-filled shell are more readily ionized than the preceding elements (see figure 12). There is a gradual increase in ionization energy from left to right across the Table in any energy sub-level of a period because of the greater coulombic attraction between the electron and the gradually increasing positive charge of the nucleus. In many vertical groups there is a decrease of ionization energy with an increase in atomic number because the large size of the atom offsets the greater positive charge, most of which is screened by the added electrons anyway. Notice, however, that this is not always true for the vertical groups of the transition elements and in particular does not hold for any of the elements from tantalum ($Z = 73$) to lead ($Z = 82$). Can you suggest a reason for this?

Electropositivity is related to ionization energy but is quite distinct from it. One measure of electropositivity is the electro-chemical series. Can you correlate this with the position of the elements in the Periodic Table? In more quantitative terms the electro-chemical series depends on the electrode potential of the metal in contact with an aqueous solution of its ions; this potential gives the tendency of the metal to go into solution. This should be compared with the ionization energy but other factors come in to the comparison so that the two properties, although broadly parallel, show considerable differences in detail.

Another measure of the electropositivity of an element is the relative basicity of its hydroxide. We are all familiar with the powerful basic character of the alkali metal hydroxides and the increase in basicity from lithium to caesium. The alkaline earths are somewhat less basic but there is an increase from beryllium and magnesium, through calcium and strontium, to barium. Successive vertical groups of the elements become progressively less basic and less electropositive until, on the righthand side of the Periodic Table, the elements are generally acidic and electronegative. In each of the groups of non-transition elements electropositivity increases with atomic number as can be seen by comparing the metallic elements thallium, lead, and bismuth with their lighter counterparts, boron, carbon, and nitrogen. Even elements such as tellurium and iodine have well developed tendencies towards electropositive behaviour despite the predominantly electronegative properties of the lighter elements in their groups. The reasons for these trends are similar to those underlying the trends in ionization energy and will become more evident when the trends in atomic size have been discussed. It is important to notice that these trends do not always carry over to the transition elements. Zinc is much more electropositive, basic, and reactive than mercury and the same can be said for iron, cobalt, nickel, and copper when compared with the platinum metals and gold. Can you suggest reasons for this?

Electronegativity of an atom is to some extent the inverse of electropositivity. It can be described as the extent to which an atom in a covalent compound attracts electrons towards itself. Rough average values of electronegativity are given in table 3. Plot these out according to atomic number, on a blank Periodic Table, draw circles having a diameter proportional to the electronegativity and note the trends along each period and down each group.

Other properties, which are perhaps more readily visualized, show a similar periodicity in their values. Thus, density, melting point, and molar heat of vaporization of the elements (table 3), all show pronounced trends and regularities which immediately become apparent when these are plotted on a graph or displayed visually as circles of appropriate diameter on a blank Periodic Table. This is well worth doing. It could also be done for other properties – ionization energies are given in figure 12 and other data on properties can be obtained from the Nuffield Advanced Science *Book of data*.

		Relative atomic mass	**Electro-negativity**	**Melting point/°C**	**Density /g cm^{-3}**	**Molar heat of vaporization /kJ mol^{-1}**
1	H	1.0	2.1	−259	0.07	0.5
2	He	4.0		−270	0.12	0.1
3	Li	6.9	1.0	181	0.53	135
4	Be	9.0	1.5	1283	1.85	295
5	B	10.8	2.0	2027	2.55	539
6	C	12.0	2.5	3800g	2.25/3.53	717
7	N	14.0	3.0	−210	0.81	2.8
8	O	16.0	3.5	−219	1.14	3.4
9	F	19.0	4.0	−220	1.11	3.3
10	Ne	20.2		−249	1.21	1.8
11	Na	23.0	0.9	98	0.97	89.0
12	Mg	24.3	1.2	650	1.74	129
13	Al	27.0	1.5	659	2.70	294
14	Si	28.1	1.8	1410	2.33	377
15	P	31.0	2.1	44.2w	2.20/1.82	30.1/12.4
16	S	32.1	2.5	119	2.07/1.96	9.6
17	Cl	35.5	3.0	−101	1.56	10.2
18	Ar	39.9		−189	1.40	6.5
19	K	39.1	0.8	63	0.86	77.5
20	Ca	40.1	1.0	850	1.55	150
21	Sc	45.0	1.3	1400	2.99	305
22	Ti	47.9	1.5	1677	4.54	429
23	V	50.9	1.6	1917	6.11	459
24	Cr	52.0	1.6	1903	7.19	349
25	Mn	54.9	1.5	1244	7.42	220
26	Fe	55.9	1.8	1539	7.86	351
27	Co	58.9	1.8	1495	8.90	382
28	Ni	57.7	1.8	1455	8.90	372
29	Cu	63.5	1.9	1083	8.94	305
30	Zn	65.4	1.6	420	7.13	115
31	Ga	69.7	1.6	30	5.91	256
32	Ge	72.6	1.8	937	5.32	334
33	As	74.9	2.0	817	5.73	130
34	Se	79.0	2.4	217	4.79	26.3
35	Br	79.9	2.8	−7	3.12	15.0

		Relative atomic mass	Electro-negativity	Melting point/°C	Density /g cm^{-3}	Molar heat of vaporization /kJ mol^{-1}
36	Kr	83.8		−157	2.16	9.0
37	Rb	85.5		39	1.53	69.2
38	Sr	87.6	0.8	770	2.58	139
39	Y	88.9	0.9	1500	4.40	393
40	Zr	91.2	1.3	1852	6.53	582
41	Nb	92.9	1.4	2497	8.55	697
42	Mo	95.9	1.6	2617	10.22	594
43	Tc	99.0	1.8	2127	11.50	577
44	Ru	101.1	1.9	2427	12.41	569
45	Rh	102.9	2.2	1966	12.41	495
46	Pd	106.4	2.2	1550	12.02	393
47	Ag	107.9	2.2	961	10.50	255
48	Cd	112.4	1.9	321	8.65	100
49	In	114.8	1.7	156	7.31	226
50	Sn	118.7	1.7	232	7.31/5.75	290
51	Sb	121.8	1.8	630	6.68	67.9
52	Te	127.6	1.9	450	6.25	50.6
53	I	126.9	2.1	114	4.94	20.9
54	Xe	131.3	2.5	−112	3.52	12.6

Table 3
Some properties of the elements in the first five periods of the Periodic Table

Densities of the elements, when plotted in this way, indicate that:
a. Within each vertical group there is a steady trend to higher density with increase in atomic number.
b. Within each horizontal period the density increases at first and then decreases, the position of the 'light metals' in Groups I and II being particularly noticeable.
c. The metals in Groups IB and IIB (for example copper $3d^{10}4s^1$ and zinc $3d^{10}4s^2$) are very much more dense than the corresponding alkali and alkali earth metals (such as potassium $3d^0 4s^1$ and calcium $3d^0 4s^2$).
d. The greatest density occurs after the $4f$ orbitals of the lanthanide elements have been filled; thus, the heavier platinum metals all have densities greater than 20 g cm^{-3}, the greatest density being found for osmium (22.48 g cm^{-3}). This element is twice as dense as lead (11.34 g cm^{-3}) which is usually considered to be a very dense metal. Some idea of the implication of these high densities can be gauged from the fact that, if a soccer ball (which normally weighs 400–450 g) were made of solid osmium, it would weigh one-eighth of a tonne: it would also cost a great deal–over £1 million.

More complicated arguments must be used to construct tables of *ionic radii*· because here there is a problem of apportioning the inter-nuclear distance between the two chemically different atoms. However, when this is done, similar trends emerge and these are basic to our understanding of periodic trends in the chemistry of the elements. Since these ideas permeate the whole of our interpretation of inorganic chemistry it is not possible to give a full treatment here, but a few examples will illustrate the great use these concepts have in correlating apparently unconnected facts within a broad and systematic framework.

1 Cations are normally smaller than anions, the only exceptions being the four largest cations (Rb^+, Cs^+, Fr^+, and Ra^{2+}) which are larger than the smallest anion, F^-. For this reason the geometrical structure of ionic crystals is dominated by the limited number of ways in which the large anions can pack around the small cations whilst maintaining overall electrical neutrality.

2 Within each vertical group the ionic radius increases with an increase in atomic number, for example,

Li^+ 60 pm, Na^+ 95 pm, K^+ 133 pm, Rb^+ 148 pm, Cs^+ 169 pm, Fr^+ 176 pm
F^- 136 pm, Cl^- 181 pm, Br^- 195 pm, I^- 216 pm
(1 pm $= 10^{-12}$ m $= 10^{-2}$ Å)

Can you suggest a reason why a cation is always smaller than an anion with the same number of electrons? The trends in ionization potential, electropositivity, and electronegativity are related in part to these trends in atomic size, elements tending to become more basic as atomic number increases within a group. Note, however, that other factors play a part, particularly in the transition element groups, where effective nuclear charge sometimes predominates. For example, with zinc, cadmium, and mercury, the increase in atomic size which tends to permit more ready ionization is not enough to counterbalance the increase in nuclear charge which tends to restrict the removal of electrons; as a result zinc is more electropositive than mercury.

3 Within each horizontal iso-electronic sequence the radius decreases rapidly with any increase in positive charge, for example,

Na^+ 95 pm, Mg^{2+} 65 pm, Al^{3+} 50 pm, Si^{4+} 41 pm

As a result, only the first two of these elements are cationic and basic, whereas aluminium and silicon are amphoteric (that is, show both basic and acidic characteristics).

4 A corollary of the preceding two trends is that, for lighter elements in the main groups, there is a 'diagonal relationship' which rationalizes the chemical similarity of an element to its neighbour diagonally below it to the right in the Periodic Table. This is because the size increase expected from trend (**2**) above is approximately cancelled by the size decrease expected from trend (**3**). For example, there are certain similarities in the chemistry of lithium and magnesium, beryllium and aluminium, and, to a lesser extent, boron and silicon.

5 Ions with a d^{10} configuration are smaller than the preceding equivalent cation having an inert gas configuration, for example, Cu^+ 96 pm, K^+ 133 pm, and Zn^{2+} 74 pm, Ca^{2+} 99 pm. This explains the higher ionization potentials and lower electropositivity of the d^{10} cations and also their tendency not to form simple ionic compounds (compare for instance CuCl and KCl, or ZnO and CaO). It also interprets the greater ease of complex formation because of the greater attractive power of the small cations for the surrounding donor groups. Many of the other differences between the coinage metals and the alkali metals stem from the same cause.

6 Successive increases in the number of electrons removed from a cation progressively decrease its radius, for example, Fe^{2+} 76 pm, Fe^{3+} 64 pm. As a result, higher oxides are usually more acidic and less ionic than lower oxides. A good example is furnished by manganese which goes from the ionic, basic MnO, through amphoteric oxides such as MnO_2 to the covalent acidic Mn_2O_7 (Mn^{2+} 80 pm, Mn^{3+} 66 pm, Mn^{4+} 54 pm, Mn^{7+} 46 pm). It is instructive to compare this trend with the one discussed in (**3**) above.

Further trends and correlations

There are numerous other trends and correlations of the same kind which have helped to deepen our understanding of the periodic nature of chemical properties. It is always worth while to seek out such systematic variations, although it is equally important to avoid glib or superficial rationalizations of the complicated chemical behaviour of the elements which often results from the subtle interplay of a variety of structural, energetic, and kinetic factors. With this proviso in mind, however, the Periodic Table, based as it is on the details of atomic structure, gives a broad and effective framework within which to make chemical experience systematic and on which to base new predictions. We have seen that three types of trend should be sought within the Table: horizontal, vertical, and, to a lesser extent, diagonal. At the most fundamental level, the number of electrons of the elements involved in compound formation and the chemical similarities of each vertical group of the Periodic Table can be appreciated, in addition to the progressive changes observed within each group and in going from one group to the next.

Periodic trends in valency – Any theory of valency must interpret both the number of electrons involved and the type of bonding it adopts in its various compounds with other elements. The approach can best be illustrated by considering the first few elements in the Periodic Table. We have already seen that hydrogen ($1s^1$) has one electron and that in helium ($1s^2$) the first quantum level is full. Let us examine the chemical implications of this in a little more detail.

The ionization energy of hydrogen is 1310 kJ mol^{-1}, that is, it requires this amount of energy to remove all the $1s^1$ electrons from one mole of hydrogen atoms; as this is greater than the amount likely to be regained by coulombic interaction with an anion, hydrogen is not expected to form simple cationic compounds of H^+. However, much of the required ionization energy can be regained, whilst still leaving the positive charge, if the now vacant 1*s* energy level can be filled by donation of a pair of electrons. Typical donors are ammonia and water molecules, and salts of $H_3N \longrightarrow H^+$ (that is NH_4^+) and of $H_2O \longrightarrow H^+$ (H_3O^+) are well known. Alternatively, hydrogen can gain an electron to give H^- thus filling the 1*s* energy level and becoming iso-electronic with helium. For example, lithium hydride, Li^+H^- is a salt-like compound which has the sodium-chloride type structure and which melts to an ionically conducting liquid. A third possibility is for hydrogen to share its electron by covalent bonding as in H_2. This, in fact, is the most frequent way hydrogen enters into chemical combination since it avoids the problems of high ionization energy in H^+ and the instability inherent in one proton looking after two electrons in H^-. In this sense the old problem of whether hydrogen is to be classified as an alkali metal or as a halogen is seen to be completely sterile; it reduces to asking whether in hydrogen the 1*s* energy level is half full or half empty: it is both. This makes hydrogen unique, whilst still resembling in some aspects of its chemical behaviour either the alkali metals or the halogens.

Helium ($1s^2$) has a filled energy level. Its ionization energy (2370 kJ mol^{-1}) is far too high to be recovered by coulombic interaction with an anion and it has no tendency to gain an electron which would have to occupy a level of considerably higher energy. Thus helium can neither lose, gain, nor share an electron; it has no tendency to combine with other elements and is referred to as being chemically inert; it has zero valency.

Lithium ($1s^2 2s^1$) can ionize much more readily than hydrogen because the 2*s* electron is, on average, further away from the positive nucleus than is the 1*s* electron in hydrogen. It, therefore, forms compounds in which it is present as singly charged positive ions and is said to be univalent. Further ionization is energetically impossible in chemical compounds. Lithium can also bond by sharing its 2*s* electron: for example, the diatomic molecule Li_2 is stable in the gas phase, and compounds such as methyllithium, CH_3Li, are well known. The similar electronic configuration of the other alkali metals, ns^1, implies a similarity in their chemistry to that of lithium, suitably modified, of course, by the trends and gradations outlined above.

Beryllium poses something of a problem. Its electronic structure ($1s^2 2s^2$) makes it appear superficially similar to helium, and the energy required to remove the outer two electrons is as high as 2654 kJ mol^{-1}. It might therefore be thought that beryllium would be an inert gas. However, the quantum rules outlined in earlier parts of this booklet indicate that there are still some unoccupied $2p$ energy sub-levels for $n = 2$. Moreover, these are energetically accessible since the configuration $1s^2 2s^1 2p^1$, which has two unpaired electrons, is only 263 kJ mol^{-1} above the ground state, $1s^2 2s^2$, which has no unpaired electrons. Provided that the energy released in forming two covalent bonds is greater than 263 kJ mol^{-1}, as it frequently is, then this 'promotion' would be energetically favourable and beryllium would be expected to be divalent. With heavier Group II metals, the larger size permits readier ionization and ionic compounds predominate. It is worth noting, however, that but for the similarity in energy of the *ns* and *np* levels, an element like magnesium would be an insulator rather than a metal – all its electrons would be paired and none would be available for conduction.

Boron ($1s^2 2s^2 2p^1$) is tervalent rather than univalent, that is it forms compounds such as BCl_3 and B_2O_3, rather than BCl and B_2O, for the same reason as discussed with beryllium. The uncoupling and promotion of a $2s$ electron to a $2p$ level permits all three unpaired electrons to partake in covalent bonding ($2s^1 2p_x^1 2p_y^1$). Likewise, the similarity in energy of the s and p sub-levels, coupled with the greater ease of ionization, ensures that the heavier Group III elements are metallic and tervalent. A further interesting point emerges here: although there is a trend to decreasing ionization energies with increasing atomic number, there is also a trend to decreasing bond energies as a result of increasing atomic size. The balance between promotion or ionization energies on the one hand, and covalent bond energies or coulombic attractive energies on the other, becomes rather finely poised. For this reason gallium can form both tervalent and univalent compounds, and the latter series of compounds becomes even more stable with indium. By the time thallium is reached the univalent state is more stable than the tervalent, although both series of compounds do exist.

Carbon and the Group IV elements behave entirely analogously. The $2s^2 2p^2$ electrons of the carbon atom can be promoted and are distributed between the four sub-levels, $2s^1 2p_x^1 2p_y^1 2p_z^1$, thus enabling each atom to form four covalent bonds. The element, in the form of diamond, is bonded in a three-dimensional lattice by strong covalent bonds and is an insulator. Again metallic properties develop with the later members of the group and the phenomenon of two stable series of compounds is apparent, particularly for tin and lead. The intermediate elements, silicon and germanium, are neither insulators like carbon, nor metals like lead, but are 'semi-conductors' in which electrons can be excited thermally, or by light, from the covalently bound state to a freer conducting state. The sophisticated and often elegant development of this property has led to the production of photo-electric cells, transistors, and many other useful devices.

Nitrogen ($2s^2 2p_x^1 2p_y^1 2p_z^1$) cannot increase the number of unpaired electrons by promotion and so has a valency of three in its simple covalent chemistry. It can, however, achieve a co-ordination number of four by donating its non-bonding pair of electrons to another atom or group as in NH_4^+ and in $H_3N \rightarrow BF_3$. With phosphorus and the heavier members of the group a further possibility arises because the third and higher quantum levels are being occupied. Thus, phosphorus ($3s^2 3p_x^1 3p_y^1 3p_z^1$) has additional vacant $3d$ levels which are energetically accessible and so, in suitable bonding environments, quinquevalent compounds involving the five energy levels, $3s^1 3p_x^1 3p_y^1 3p_z^1 3d^1$, can be obtained; a typical example is the oxidation of PF_3 to PF_5.

Trends in Group VI can be interpreted similarly. The group valency is 2 and oxygen ($2s^2 2p_x^2 2p_y^1 2p_z^1$) is restricted to this though its coordination number can rise to 3 (as in H_3O^+) and occasionally to 4 (as in basic beryllium acetate). Later members of the group are also frequently divalent (for example, S^{2-}, SCl_2) but can also involve increasing numbers of *d* orbitals thereby producing compounds such as SF_4 and SF_6.

The stepwise increase of valency by successive units of two is a noticeable feature of all main-group elements with unpaired *p* electrons in the second and third quantum shells. It is most fully developed in the halogens (except for fluorine which can only be univalent since it has no low-lying energy levels above the $2s^2 2p^5$ ground state). The oxyanions of chlorine form a good example in which the univalence of chlorine in the hypochlorite ion (ClO^-) is increased to 3 in ClO_2^-, to 5 in ClO_3^-, and finally to 7 in ClO_4^-.

With transition elements, valency changes involve *d* electrons rather than *p* electrons and the observed behaviour is rather different. Instead of changing by two units at a time, successive valencies differing by only one unit are the rule. This is because the differences between successive ionization energies within the transition elements are much smaller than for the main group elements. An indication of this has already been seen in the plot of the first-stage ionization energies of the elements in figure 9. It therefore becomes easier to involve successive numbers of electrons either in ionization or in covalent bonding. One of the largest ranges of compounds is shown by manganese; it has members in at least nine oxidation states from -1 to $+7$. This illustrates the dangers of tying elementary theories of compound formation too closely to the electronic configuration of atoms. There *is* a relation between the two, but it is by no means simple. Enough has been said, however, to indicate the great power of the Periodic Table in systematizing, interpreting, and predicting the physical and chemical properties of the elements and their compounds. The Table is based firmly on the details of the electronic structure of the atoms and, in mirroring this structure, it gives cohesion and significance to the whole of chemistry.

period ↓ / group	IA	IIA	IIIB	IVB	VB	VIB	VIIB	VIIIB	VIIIB
1	**H** 1 $1s^1$ 1310								
2	**Li** 3 $1s^2 2s^1$ 520	**Be** 4 $1s^2 2s^2$ 900							
3	**Na** 11 $3s^1$ 500	**Mg** 12 $3s^2$ 740							
4	**K** 19 $4s^1$ 420	**Ca** 20 $4s^2$ 623	**Sc** 21 $3d^1 4s^2$ 630	**Ti** 22 $3d^2 4s^2$ 660	**V** 23 $3d^3 4s^2$ 650	**Cr** 24 $3d^5 4s^1$ 650	**Mn** 25 $3d^5 4s^2$ 720	**Fe** 26 $3d^6 4s^2$ 760	**Co** 27 $3d^7 4s^2$ 760
5	**Rb** 37 $5s^1$ 400	**Sr** 38 $5s^2$ 550	**Y** 39 $4d^1 5s^2$ 620	**Zr** 40 $4d^2 5s^2$ 660	**Nb** 41 $4d^4 5s^1$ 660	**Mo** 42 $4d^5 5s^1$ 680	**Tc** 43 $4d^5 5s^2$ 700	**Ru** 44 $4d^7 5s^1$ 710	**Rh** 45 $4d^8 5s^1$ 720
6	**Cs** 55 $6s^1$ 380	**Ba** 56 $6s^2$ 500	**La** 57 $5d^1 6s^2$ 540	**Hf** 72 $4f^{14} 5d^2$ $6s^2$ 680	**Ta** 73 $4f^{14} 5d^3$ $6s^2$ 760	**W** 74 $4f^{14} 5d^4$ $6s^2$ 770	**Re** 75 $4f^{14} 5d^5$ $6s^2$ 760	**Os** 76 $4f^{14} 5d^6$ $6s^2$ 840	**Ir** 7 $4f^{14} 5d^7$ $6s^2$ 880
7	**Fr** 87 $7s^1$ 380	**Ra** 88 $7s^2$ 510	**Ac** 89 $6d^1 7s^2$ 660						

period					
6	**Ce** 58 $4f^2 5d^0$ $6s^2$ 670	**Pr** 59 $4f^3 5d^0$ $6s^2$ 550	**Nd** 60 $4f^4 5d^0$ $6s^2$ 610	**Pm** 61 $4f^5 5d^0$ $6s^2$	**Sm** 6 $4f^6 5d^0$ $6s^2$ 540
7	**Th** 90 $5f^0 6d^2$ $7s^2$ 670	**Pa** 91 $5f^2 6d^1$ $7s^2$	**U** 92 $5f^3 6d^1$ $7s^2$ 390	**Np** 93 $5f^4 6d^1$ $7s^2$	**Pu** 9 $5f^6 6d$ $7s^2$

Figure 12
Periodic Table with atomic numbers, electron structures, and ionization energies (kJ mol^{-1}).

VIIIB	IB	IIB	IIIA	IVA	VA	VIA	VIIA	inert gases
								He 2 $1s^2$ 2370
			B 5 $1s^22s^2$ $2p^1$ 800	**C** 6 $1s^22s^2$ $2p^2$ 1090	**N** 7 $1s^22s^2$ $2p^3$ 1400	**O** 8 $1s^22s^2$ $2p^4$ 1310	**F** 9 $1s^22s^2$ $2p^5$ 1680	**Ne** 10 $1s^22s^2$ $2p^6$ 2080
			Al 13 $3s^23p^1$ 580	**Si** 14 $3s^23p^2$ 790	**P** 15 $3s^23p^3$ 1010	**S** 16 $3s^23p^4$ 1000	**Cl** 17 $3s^23p^5$ 1260	**Ar** 18 $3s^23p^6$ 1520
Ni 28 $3d^84s^2$ 740	**Cu** 29 $3d^{10}4s^1$ 750	**Zn** 30 $3d^{10}4s^2$ 910	**Ga** 31 $3d^{10}4s^2$ $4p^1$ 580	**Ge** 32 $3d^{10}4s^2$ $4p^2$ 760	**As** 33 $3d^{10}4s^2$ $4p^3$ 950	**Se** 34 $3d^{10}4s^2$ $4p^4$ 940	**Br** 35 $3d^{10}4s^2$ $4p^5$ 1140	**Kr** 36 $3d^{10}4s^2$ $4p^6$ 1350
Pd 46 $4d^{10}5s^0$ 800	**Ag** 47 $4d^{10}5s^1$ 730	**Cd** 48 $4d^{10}5s^2$ 870	**In** 49 $4d^{10}5s^2$ $5p^1$ 560	**Sn** 50 $4d^{10}5s^2$ $5p^2$ 710	**Sb** 51 $4d^{10}5s^2$ $5p^3$ 830	**Te** 52 $4d^{10}5s^2$ $5p^4$ 870	**I** 53 $4d^{10}5s^2$ $5p^5$ 1010	**Xe** 54 $4d^{10}5s^2$ $5p^6$ 1170
Pt 78 $4f^{14}5d^9$ $6s^1$ 860	**Au** 79 $4f^{14}5d^{10}$ $6s^1$ 890	**Hg** 80 $4f^{14}5d^{10}$ $6s^2$ 1010	**Tl** 81 $4f^{14}5d^{10}$ $6s^26p^1$ 590	**Pb** 82 $4f^{14}5d^{10}$ $6s^26p^2$ 720	**Bi** 83 $4f^{14}5d^{10}$ $6s^26p^3$ 700	**Po** 84 $4f^{14}5d^{10}$ $6s^26p^4$ 810	**At** 85 $4f^{14}5d^{10}$ $6s^26p^5$	**Rn** 86 $4f^{14}5d^{10}$ $6s^26p^6$ 1040

Eu 63 $4f^75d^0$ $6s^2$ 550	**Gd** 64 $4f^75d^1$ $6s^2$ 590	**Tb** 65 $4f^95d^0$ $6s^2$ 650	**Dy** 66 $4f^{10}5d^0$ $6s^2$ 660	**Ho** 67 $4f^{11}5d^0$ $6s^2$	**Er** 68 $4f^{12}5d^0$ $6s^2$ 590	**Tm** 69 $4f^{13}5d^0$ $6s^2$ 590	**Yb** 70 $4f^{14}5d^0$ $6s^2$ 600	**Lu** 71 $4f^{14}5d^1$ $6s^2$ 590

Am 95 $5f^76d^0$ $7s^2$	**Cm** 96 $5f^76d^1$ $7s^2$	**Bk** 97 $5f^96d^0$ $7s^2$	**Cf** 98 $5f^{10}6d^0$ $7s^2$	**Es** 99 $5f^{11}6d^0$ $7s^2$	**Fm** 100 $5f^{12}6d^0$ $7s^2$	**Md** 101 $5f^{13}6d^0$ $7s^2$	**No** 102 $5f^{14}6d^0$ $7s^2$	**Lr** 103 $5f^{14}6d^1$ $7s^2$